Future Energy

Wind Power

Julie Richards

This edition first published in 2004 in the United States of America by
Smart Apple Media.

All rights reserved. No part of this book may be reproduced in any form or
by any means without written permission from the publisher.

Smart Apple Media
1980 Lookout Drive
North Mankato
Minnesota 56003

Library of Congress Cataloging-in-Publication Data
Richards, Julie.
  Wind power / Julie Richards.
  p. cm. — (Future energy)
  Summary: Describes how wind can be used as a source of power, how it affects the
  environment, and what the significance of wind power may be in the future.
  ISBN 1-58340-333-7
  1. Wind power—Juvenile literature. [1. Wind power.] I. Title.
  TJ820.R52 2003
  333.9'2—dc21   2002044638

First Edition
9 8 7 6 5 4 3 2 1

First published in 2003 by
MACMILLAN EDUCATION AUSTRALIA PTY LTD
627 Chapel Street, South Yarra, Australia 3141

Associated companies and representatives throughout the world.

Copyright © Julie Richards 2003

Edited by Anna Fern
Text and cover design by Cristina Neri, Canary Graphic Design
Illustrations by Nives Porcellato and Andy Craig
Photo research by Legend Images

Printed in Thailand

Acknowledgements
The author and the publisher are grateful to the following for permission to
reproduce copyright material:

Cover photograph: a wind farm, courtesy of Getty Images.

Australian Picture Library/Corbis, pp. 8, 9, 16, 19, 23, 25, 27; Bureau of Meteorology,
p. 5 (bottom); Coo-ee Picture Library, pp. 6 (top), 15, 24; East Bay Newspapers, p. 21;
Getty Images, pp. 1, 5 (center), 7, 12, 17, 18, 20, 22, 26, 30; Nasa, p. 29; NEG Micon,
p. 28; Photolibrary.com, pp. 4, 5 (top), 13 (bottom), 14 (top).

While every care has been taken to trace and acknowledge copyright, the publisher
tenders their apologies for any accidental infringement where copyright has proved
untraceable. Where the attempt has been unsuccessful, the publisher welcomes
information that would redress the situation.

Contents

What is energy?	4
The wind as a source of energy	5
Where does wind energy come from?	6
Using wind energy	7
Wind energy through history	8
Early wind technology	10
Modern wind technology	11
Harnessing wind energy	12
Wind energy and the environment	17
Wind power at work	18
Wind power in the future	28
Advantages and disadvantages of wind power	30
Glossary	31
Index	32

Glossary words
When a word is printed in **bold** you can look up its meaning in the glossary on page 31.

What is energy?

Energy makes the world work. People, plants, and animals need energy to live and grow. Most of the world's machines are powered by energy that comes from burning coal, oil, and gas. Coal, oil, and gas are known as fossil fuels. Burning fossil fuels makes the air dirty. This is harmful to people and damages the environment.

Scientists are not sure how much longer fossil fuels will last. It depends on whether or not new sources of this type of energy are found and how carefully we use what is left. Scientists do know that if we keep using fossil fuels as fast as we are now, they *will* run out. An energy source that can be used up is called non-renewable. A renewable source is one that never runs out. The world cannot rely on fossil fuels as a source of energy for everything. We need to find other sources of safe, clean, renewable energy to power the machines we have come to depend on.

These are the fossilized remains of a fish. Fossil fuels are the plants and animals that died millions of years ago and turned into coal, oil, and gas.

Wind Power

The wind as a source of energy

The wind gets its energy from movement. Wind is air moving across the surface of the Earth. We cannot see the wind because air is invisible. When the wind is blowing, we can see what it does or where it has been. Sometimes, the wind blows so gently that it hardly moves the leaves on the trees. At other times, winds blow so strongly that they tear trees from the ground and topple buildings. Wind also makes waves when it blows across the surface of water. Strong winds can whip up huge waves that sink ships and flood nearby land. Winds can lift the soil from the land and carry it many miles away. This means there is less soil on the ground for next year's **crops** to grow in. Even the desert sand is driven into huge piles by the wind and pushed, grain by grain, around the desert.

The wind is a very powerful force of nature. The energy from its movement can be **harnessed** to do useful work. The wind will never run out. It is a source of clean, renewable energy.

A gentle breeze is not strong enough to move this flag.

These powerlines were blown down by strong winds.

This dust storm lifted 220,400 tons (200,000 t) of soil and carried it to the city of Melbourne, Australia.

Fact file
Antarctica is the windiest continent in the world. Winds can blow at speeds of more than 125 miles (200 km) per hour.

Where does wind energy come from?

All winds are made by the Sun. Each day the ground soaks up warmth from the sunshine. When this happens, the air just above the ground is warmed. Air that is warmed becomes lighter and begins to rise. Cooler, heavier air rushes in to fill the space the warm air leaves behind. When cool air and warm air move about like this, we call it wind.

Some parts of the Earth warm up more quickly than others. A shady forest will take a lot longer to heat up than a desert because the trees block some of the sunlight. Land heats up faster than the sea because the water is always moving and carrying the heat away. Because these different surfaces heat at different speeds, the air is also moving at different speeds. While a gentle breeze blows in one place, a fierce wind might be raging a few hundred miles away.

A windsock is a tube of fabric attached to a pole. Windsocks show us how fast and which way the wind is blowing.

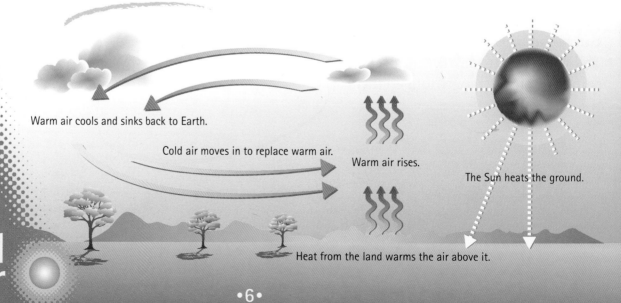

When warm air and cool air move around, we call it wind.

Warm air cools and sinks back to Earth.

Cold air moves in to replace warm air.

Warm air rises.

The Sun heats the ground.

Heat from the land warms the air above it.

Wind Power

Using wind energy

Natural wind energy is used to dry clothes and move sailing boats and windsurfers about. It is used to pump water from the ground using a windmill and to fly kites and gliders.

Hundreds of years ago, sailors relied on natural wind energy to push their boats across vast oceans. If the wind did not blow, they did not move. Sometimes they were stuck for many days or weeks and died of thirst or starvation. Winds that were too strong would push their boats onto jagged rocks and sink them. The wind was also used to power simple machines that could grind corn or raise water from **wells**. Although the wind is a very useful source of energy, it cannot be switched on or off as needed. People have always searched for more reliable sources of energy.

During the 1800s, a source of reliable energy became available. It was called electricity. Electricity is made by burning fossil fuels inside a power station. The electricity is sent along wires to wherever the machines are. Electricity is a source of energy that will never run out. However, the fossil fuels that are burned to make electricity are beginning to run low. Burning them is damaging the environment. As fossil fuels give up their energy during burning, dangerous gases and chemicals escape into the air. These gases and chemicals pollute the air we breathe. **Acid** rain is made when chemicals mix with water in the air. As it falls, it slowly eats away at anything made of stone or wood and poisons forests and rivers.

Natural wind energy is used for simple, everyday things.

Wind energy through history

Wind is a natural energy that people have used for thousands of years in transportation and to run simple machines.

Ancient use of wind energy

The first sailing boats were used by the Egyptians, about 4,000 to 5,000 years ago. Early sailing boats had square sails and could only travel in the direction the wind pushed them. In the 1100s, Arab sailors invented a triangular sail. This new sail allowed them to travel in any direction they wanted. Since that time, people have used sailing ships and boats for fishing and to explore other parts of the world and **trade** with the people they met. Many famous battles were fought at sea in sailing ships. Shipbuilding created jobs. Towns and cities grew up around the shipyards alongside the rivers and coasts of many countries.

About 3,000 years ago, in China, large, strong war kites were used to carry weapons and food to soldiers surrounded by the enemy. Other kites pulled wheeled carts across open land, where the wind blows strongest. It was faster and safer than relying on animals that needed food and water or became tired too easily.

Fact file
For more than 50,000 years, people have used fire to clear land. Once the fire was lit, the people used their knowledge of the wind to control it.

An illustration from the 1200s showing an Arab merchant's sailing dhow.

Windmills like this are still used throughout Greece to pump water from underground.

Modern use of wind energy

Modern people use wind energy in many different ways. Some of the first machines invented were powered by wind energy. The first windmills were built in places such as Iran and Afghanistan. Soldiers returning to Europe from wars in these countries told stories about the windmills they had seen. Windmills became very popular because, unlike machines that relied on water power, windmills did not have to be built near running water. In 1712, the first steam-powered engine was invented by Thomas Newcomen. Steam power was to change the world.

Sail power to steam power

Sailing ships began to disappear. Ships with steam-powered engines were faster, stronger, and more reliable than ships that used sails. Coal was used as fuel to heat the water to steam because it burned longer and hotter than wood.

Steam power is made by heating water inside a boiler until it turns to steam. Blasts of steam are used to drive engines and wheels in machines. The machines can do the work of many people much faster. Steam power was useful for transportation.

During the 1800s, when more people began using electricity, even more coal had to be dug up and burned. Even though natural wind energy was still useful, some inventors began to think of ways to harness wind energy and change it into a more reliable source of energy such as electricity.

Early wind technology

The windmill was the earliest form of wind technology. Windmills did not always make the best use of natural wind energy and, over time, their design changed.

Early windmills

The first windmills looked very different from the windmills we know today. Early windmills were **horizontal**, with wooden sails that turned like helicopter rotor blades. The sails were attached to a post that turned a wheel. As the wheel turned, it made two flat, heavy stones rub together. Grain was poured in between the stones and, as the stones rubbed together, the grain was crushed into flour.

Post windmills

Inventors discovered that windmills were more powerful if the sails were placed in an upright position. These windmills had sails made of cloth and hung from an upright wooden post. The post was turned around so that the sails faced into the wind. Every time the wind changed, the post had to be turned by hand, or the sails would not catch the best wind.

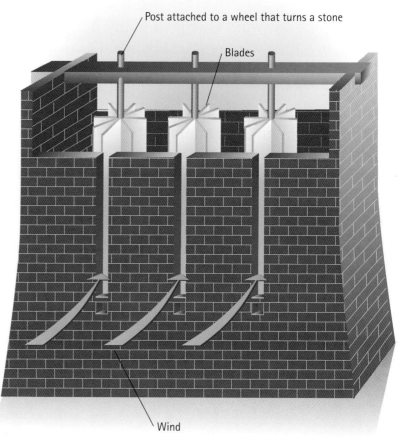

This type of early windmill was used to grind grain. It was not very powerful.

Modern wind technology

It was soon discovered that wind-powered machines could be made more powerful by changing the way the wind energy was collected and used.

Tower windmills

The tower windmill was the most powerful of all the windmills. The stone-tower windmill did not turn like the old post windmills. Only the top part of the windmill, where the sails were, turned. A wheel at the back of the tower, called a fantail, kept the sails pointed in the right direction at all times. When the wind blew the fantail around, it turned the sails into the wind. When the fantail was pointing in the same direction as the wind, it would stop, holding the sails in the best position for gathering the wind's energy.

The sails had shutters in them to control the speed of the windmill. When the shutters were opened, the wind would blow through the sails, so they went more slowly. When the shutters were closed, the wind pressed against the sails harder and pushed them around faster. Tower windmills were used to grind grain and pump water.

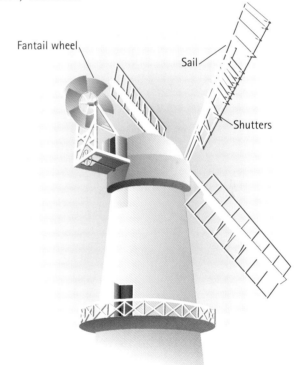

Electricity from windmills

Windmills can be used as a source of energy to make electricity. We depend upon electricity to run the millions of machines in the world. We are burning fossil fuels so fast to make enough electricity that they will soon run out. Fossil fuels are harmful to the environment and people's health because they cause pollution when they are burned. The wind is a clean and safe source of renewable energy that can never run out. Scientists have found ways to make windmills that are powerful enough to generate electricity.

The fantail wheel was invented in the 1700s. This wheel kept the sails facing into the wind at all times without anyone having to turn them by hand.

Harnessing wind energy

The clean, renewable energy of the wind has many uses in everyday life.

Turning wind into electricity

The modern windmills that are used to make electricity today are called wind **turbines**. Windmills have been used to make electricity for more than 100 years.

Older-style windmills

From the 1800s, people on farms in **remote** areas of the United States and Australia were able to make their own electricity by connecting a windmill to a **generator**. The generator was a **magnet** inside a wire coil which was fitted behind the windmill blades. When the wind turned the blades, the magnet would spin inside the wire coil. The energy from this spinning movement was changed into electrical energy and sent through the wire. This was a very cheap way to make electricity. Unfortunately, the electricity only lasted as long as the wind was blowing. These windmills had many blades, so they could catch as much wind energy as possible. They also had a fantail that kept them facing into the wind.

This windmill makes electricity for a small household in Ireland. When the wind spins the blades, the wind energy is changed into electricity.

Modern wind turbines

A modern wind turbine has a steel framework or hollow concrete tower which raises the turbine high into the air where much stronger winds blow. The tower has a set of blades which are specially curved to catch the wind. Some wind-turbine blades look a bit like giant propellers. Others are like enormous egg-beaters! The difference between them is the direction in which the blades turn. There are two types of modern wind turbine: horizontal-**axis** and **vertical**-axis.

Horizontal-axis wind turbines

A horizontal-axis wind turbine is a tower with blades that look like propellers fitted to it. Most modern turbines have two or three of these blades, which spin just like the propellers on an aircraft.

Wind turbines 4, 5, and 6 catch the wind much more easily than 1, 2, or 3.

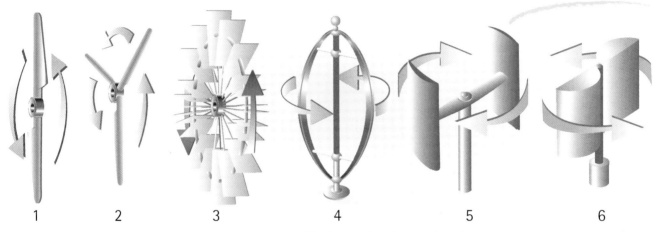

1 2 3 4 5 6

Horizontal-axis wind turbines

Horizontal-axis wind turbines do not catch the wind very easily. A computer-controlled motor inside the top part of them keeps the blades facing into the wind so the generator can keep making electricity. Unfortunately, this type of wind turbine has its generator at the top of the tower as well. Workers must climb the tower whenever they need to check or repair the generator. The tower has to be very strong to support the heavy weight of the equipment and the turbine above it.

Vertical-axis wind turbines

Vertical-axis wind turbines have large arms that spread out from the tower. The blades catch the wind in a different way to horizontal-axis turbines. One type of vertical-axis wind turbine is called the Darrieus rotor. The Darrieus rotor is named after the Frenchman George Darrieus, who invented it in the 1930s. A Darrieus rotor looks very much like a giant egg-beater. A vertical-axis turbine works well whichever direction the wind blows in. A vertical-axis turbine has all of the heavy equipment on the ground, so the tower can be thinner and lighter.

Darrieus rotor wind turbines

Inside a wind turbine

At the top of a wind turbine, behind the blades, is a control room, the **nacelle**, where the generator and the motor sit. The motor turns the turbine into the wind so the blades keep spinning. The generator changes the spinning movement of the blades into electricity. As high winds can damage the blades, most wind turbines have a special control that slows down the blades when the wind reaches a certain speed.

How are wind turbines controlled?

Wind turbines are controlled by computers that track the wind's direction and measure its speed. The direction and speed of the turbine blades are adjusted to suit the wind. If the wind blows too strongly, the blades can be turned out of the wind so they cannot spin too quickly. Large wind turbines have brakes that slow the blades down. Smaller turbines use weights attached to their blades to slow their speed.

Wind turbines can be as tall as a 25-story building. The part containing the control box can weigh up to 25 tons (23 t).

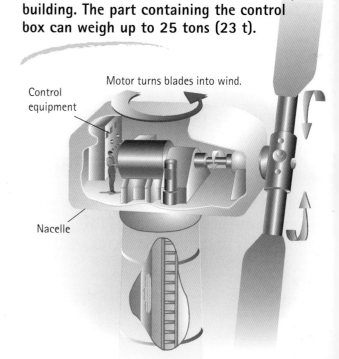

Motor turns blades into wind.
Control equipment
Nacelle

Wind farms

Lots of wind turbines can be put together to make a wind farm. The electricity is passed through a special machine that changes it into the type of electricity we use in our homes. It is then sent through overhead powerlines to nearby towns and cities. The electricity can also be stored in **batteries** for when it is needed. Electricity made this way means that no fossil fuels need to be burned. Wind energy does not produce any dangerous gases or harmful chemicals. It does not use up a valuable source of energy that cannot be renewed.

Fact file
In 1941, an American named Palmer Puttnam built the world's first really big wind turbine on top of a mountain. The turbine had two blades. Each was 175 feet (53 m) from tip to tip and sat on top of a 109-foot (33-m) tower. It worked successfully until March 1945, when one of its massive blades flew off.

Wind farms can be built quickly, and more turbines can be added as they are needed. Farmers who have sheep and cattle can keep their animals on the same land and make a living from farming the wind as well. Wind farms are usually built on open land or on the sides of hills, where the wind is likely to be blowing most of the time. The turbines need to be spaced well apart. Their towers have to be sunk up to 165 feet (50 m) into the ground. Explosives are sometimes used to break up rocky soil. A tall crane is needed to lift the turbine into position. Wind farms can be quite expensive to set up. The technology is becoming cheaper, as scientists find ways to build better and less expensive wind turbines. Although the wind-turbine towers are slim, wind farms do take up a lot of space that could be used for other things.

There are more than 40,000 wind turbines in the world. California, in the U.S., has 17,000 turbines. The turbine in this photo is in Australia.

In very dry countries, water can be close to the Earth's surface, but difficult to reach. Wind energy can drive a pump that raises the water.

Small wind turbines for remote areas

Wind turbines come in different sizes. Small wind turbines are perfect for people living in remote areas and **developing countries** where there is no established electricity supply. Small wind turbines come in kits. The blades are usually made from **fiberglass** or aluminum. The tower is made of a lightweight steel. The electricity is carried down to the ground by wires inside the tower.

Small wind turbines are easy to set up and are useful for the smaller energy needs of remote villages, such as pumping water. As there are no powerlines to carry electricity, power can be stored in batteries for later use. People in local communities can be trained in how to build, use, and repair the turbines.

Do-it-yourself wind turbines

Some people have made their own wind turbines using simple everyday things such as a bicycle wheel or an oil drum cut in half. The two halves of an oil drum are joined onto a steel pipe. The pipe is then slipped over a pole so that it can spin. The wind always fills the open half of the oil drum and pushes it around. The generator is often taken from an old car that is no longer used.

Wind energy and the environment

Wind energy is a clean, renewable source of energy. It does not pollute the environment by releasing dangerous gases and chemicals. It does not make acid rain or add to **global warming**. So long as the Sun shines, wind will always be there as an environmentally friendly choice of energy.

However, some people are worried about wind turbines. They think that wind turbines are ugly and that wind farms spoil the beauty of the countryside. Most wind-turbine towers are made of white concrete. Some towers have been made of a steel framework that looks like a crane. Because they are "see-through," they blend in better with the countryside around them. Other people feel that the constant swish of wind-turbine blades is too noisy. Scientists have developed new wind turbines with much quieter generators and less blade noise. Some **environmentalists** believe wind farms endanger wildlife. Blasting disturbs the ground and may drive some animals away forever. Eagles hunting for food have been known to fly into the spinning blades.

When a wind farm is no longer needed, the turbines are taken apart and the pieces removed. The concrete bases are usually left in the ground and covered with soil. The operators of the wind farm must leave the land as close as possible to the way it was before the wind farm was built. They can only use the same types of soil, rock, and plants that would normally live there.

Fact file
The blades on a wind turbine are painted light gray because this color blends best with any surroundings.

Towers like this blend in better with the countryside.

Wind power at work

Wind power is useful for things such as pumping water. Wind power is also being used once again as a source of clean energy for sea transportation. Natural wind energy powers many of the sports and activities people enjoy.

Pumping water

Most wind turbines are used to pump water. In many parts of the world, it is difficult to grow crops all through the year because of the weather. Sometimes, the water is so deep underground that grasses cannot grow and sheep and cattle do not get enough to eat. Before wind turbines were invented, water pumps were driven by burning a fossil fuel called kerosene. Even though kerosene was cheap to buy, the pumps were not very **efficient** and often needed repairing. Although wind turbines cost more to set up, they are more reliable. They are cheaper and cleaner to run because they do not burn any fuel.

Fact file
The world's largest wind turbine is in Oahu, Hawaii. Its blades are 320 feet (97.5 m) across.

Wind-powered pumps bring drinking water to the surface in places where the land and the weather are very dry.

This passenger cruise ship has sails as well as engines. Using the wind cuts costs and helps prevent pollution because less fuel is burned.

Wind-powered transportation

Thousands of years ago, people discovered that wind-powered boats and kites were faster and more reliable than using animals as transportation. Modern scientists and inventors have found even better ways to use the wind. Some wind-powered transportation uses natural wind energy. Other types make their own wind.

Using natural wind energy in sea transportation

Hundreds of years ago, early inventors had found that by changing the shapes of sails, adding extra sails, and building high masts, the wind could push their ships along even faster. Even so, sailing ships would never be as fast or reliable as the new ships with their big powerful engines. The last big sailing ships were used in 1914 to carry fertilizer from South America to Europe.

Modern sailing ships

Today, the wind is still used as a source of energy to drive some passenger liners, oil-tankers, and cargo ships. These ships still have engines powered by fossil fuels, but, when there is a good wind blowing, they use sails instead. This helps cut pollution because less fuel is burned. The world's first wind- and solar-powered ferry, *Sydney Solar Sailor*, was launched in February 2000, in Australia. *Sydney Solar Sailor* has large, curved solar panels attached to masts. These solar panels catch the wind just as sails do.

> **Fact file**
> During the 1850s, sailing ships called clippers carried wool from Australia to Britain. They had huge sails which made them very fast. The name "clipper" came from the way they could "clip" time off the journey.

These yachts are using the wind in their sails to push them across the water.

Yachts

A yacht is a sailing boat that uses the wind's energy to move it across the water's surface. The force of the wind hitting the sail pushes the sail and the boat forward in one direction. To be able to sail in any direction, it is necessary to turn the sail to catch the wind. A **rudder** is also used to turn the boat.

Airfoils

An airfoil is something that shapes the wind around it. A sail on a yacht is a type of airfoil. Kites, hang-gliders, windsurfers, and wind-skaters are also airfoils.

Fact file

Nature has always used the wind as transportation. Some seeds have wings which act as airfoils, enabling the seeds to travel up to 1.2 miles (2 km) from the tree they fell from. The wind also carries smells long distances. By smelling the wind, insects and animals can find food and avoid danger.

How an airfoil works

An airfoil has a special curved shape that makes the air flow past or over it in a certain way. As the airfoil moves through the air, some of the air pushing past it becomes thinner and weaker. The stronger air pushes against the thinner, weaker air pushing the airfoil forward.

Ships that use airfoils

Some modern ships are fitted with wing sails. Wing sails are a type of airfoil. Wing sails stand upright and make use of winds that blow far out to sea. Wing sails are curved, and the thin end of the sail faces the wind. When the wind blows over the surface, the thinner, weaker air at the wider end of the sail is pushed away by stronger air. This pulls the ship forwards.

Computers are used to track the speed and direction of the wind. They control the wing sails and move them so that they catch the best wind.

Wing sails on a modern ship

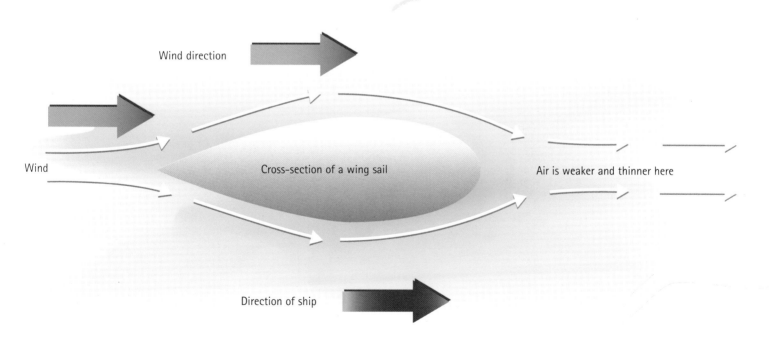

An airfoil is something that shapes the wind around it.

Wind-powered flight

Wind-powered flight is all about using the wind to keep an aircraft in the air and to control the aircraft's movements. Aircraft that use natural wind energy in flight still need help getting **airborne**.

Gliders

A glider is an aircraft that uses the power of the wind to stay airborne. A glider cannot take off by itself because it has no engine. Gliders have to be towed into the air by other aircraft. The tow-rope is then taken away and the glider uses the wind and rising warm air from the ground to keep flying.

Hang-gliders

A hang-glider is like a giant sail. The pilot is held beneath the sail in a harness and steers the kite by pulling on a control bar. Hang-gliders have to be launched from a high cliff top or hill, where there is a lot of wind rushing upwards. The pilot runs into the wind and points the hang-glider into the air. Because a hang-glider is an airfoil, it uses its shape to split the air and lift it into flight.

A balloon uses air heated by gas burners to lift it from the ground. The pilot uses the wind to steer the balloon and can use the winds at different heights to control the speed of the balloon's flight.

Fact file

In 1853, a hang-glider built by George Cayley carried a person for the first time. A German man named Otto Lilienthal made more than 200 flights in his homemade hang-glider, which he launched from a specially built hill. Unfortunately, he was killed in 1896 when his hang-glider crashed.

A quadrifoil is a kite that uses the wind to pull a skier or a sled across ice and snow.

Wind-powered adventure

Kites and sails can also be used as land transportation.

Quadrifoils

A quadrifoil is a land kite specially designed for traveling across snow and ice. Sled dogs are no longer allowed in Antarctica. Adventurers exploring Antarctica on skis now use quadrifoils to pull their sleds instead. A quadrifoil is attached to a sled by strong ropes that cannot freeze or split in the harsh Antarctic weather. The kite is made up of lots of pockets of **synthetic** material filled with air. The skier wears a harness connected to the sled. The strings that connect the harness to the sled allow the skier to steer the quadrifoil by pulling them in different directions. The strings are made from a special rope that becomes stiff in very cold weather. This stiffness stops the ropes from knotting and the kite becoming tangled. Although the skier is in front of the quadrifoil, it is the wind that moves it along. This means the skier does not become too tired too quickly. Steering the quadrifoil gives the skier enough exercise to stay warm.

Air-cushion vehicles

Air-cushion vehicles (ACVs) are designed to travel on land and water. They are also called hovercraft. An ACV rides on a cushion of air created by fans inside its body. The fans, which work like the rotor blades on a helicopter, suck air into the space underneath the vehicle and trap it behind a rubber skirt. When this space is filled with air, the ACV is lifted and hovers just above the ground on the cushion of air it has created. Another fan or propeller is mounted on the back, to drive the vehicle forward. Because an ACV sits on air, it can travel easily over water or land. The rubber skirt is **flexible**. It bends easily when the vehicle moves over uneven ground.

Because ACVs travel on a cushion of air, they are very quiet. Being able to ride from water to land, they have been used as rescue vehicles or to bring soldiers from ships to the shore. Unfortunately, ACVs still use fossil fuels to power their motors. It is possible that **fuel cells** and new cleaner liquid fuels and gases might be used in ACVs. But ACVs have heavy engines and fuel tanks, which need to be supported while the vehicle is hovering. So far, only leaded **gasoline** can do this. If laws are passed to stop people using leaded gasoline, ACVs may be a thing of the past unless somebody can work out a better source of fuel. Environmentalists are keen to keep them, because they can travel through delicate wetlands without damaging the plants or harming wildlife.

An air-cushion vehicle is also called a hovercraft. It travels easily over water or land.

Wind Power

Kite festivals are very popular in Japan and Southeast Asia. This huge kite was photographed in Tokyo, Japan, at an annual giant kite-flying event.

Wind power for fun

People can enjoy many thrilling sports that use the natural energy of the wind.

Kites

Kites are the oldest flying machines. Kite-fliers control their kites by pulling on a string. A tail at the back of the kite adds enough extra weight to keep the kite steady. By pulling the string, the kite-flier controls the way the wind hits the kite, so that it catches the most wind possible. During the Boer War in South Africa, in 1899, giant kites were used to carry spies over enemy camps. Before weather balloons were invented, kites were launched to get information on the wind's speed and direction. Today, many Chinese farmers still fly kites to help them work out what sort of weather may be coming.

Kite fighting is a popular sport in Japan and Southeast Asia. Many of the brightly colored kites are made in the shapes of birds, dragons, and fish. Some of them have pipes and whistles that make musical sounds as the wind blows through them. These kites can be enormous, and several people might be needed to control them. The kites can be flown at heights of 2,000 to 3,000 feet (600–900 m). The kites are used as weapons. Their strings are coated in powdered glass and sharp blades are embedded in their framework. Each team uses their kite to attack the other team's kite and bring it down.

Wind-skating

A wind-skater uses the wind in much the same way as a hang-glider. The wind-skater holds an A-frame sail above their head and uses skates to build up speed at ground level. By holding the sail at an angle, the sail becomes an airfoil. The wind above the sail is thinner and weaker. This makes the wind under the sail press upwards. The wind-skater uses the weight of their body to control the angle of the sail and the direction of the flight. Unlike hang-gliders, wind-skaters make short flights. A lot of effort goes into performing tricks and jumps in the air.

Windsurfing

A windsurfer uses a surfboard with a triangular sail attached to it to move across the water's surface. A windsurfer uses the power of the wind in the same way a yacht does. The wind fills the sail, pushing the board forwards. The windsurfer has a flat board underneath it called a dagger board. The dagger board stops the windsurfer from being blown over sideways.

A windsurfer

Fact file

The ancient Greek engineer Hero of Alexandria designed a musical organ that was powered by the wind. Hero used the energy of the wind to push a rod up and down forcing air through the organ pipes. As the air around the pipes vibrated, it made a sound. Most modern pipe organs still work this way.

Preparing to race a land-yacht. One man steers, and the other balances the top of the mast.

Land-yachts and ice-yachts

Land-yachts look very much like go-carts with sails. The pilots lie back in their seats to allow the wind to flow over their bodies. If they sat up, their bodies would stop the wind from moving over them as fast. This would slow the land-yacht down. Land-yachts usually race along beaches, because the wind blowing off the sea is more powerful than the wind blowing across the land.

Ice-yachts skim across frozen surfaces on three sharp, stainless-steel runners. The pilot lies down inside what looks like the body of a Formula-One racing car. Ice-yachts are steered by turning the front runner.

Wind power in the future

The wind is a source of clean, safe, renewable energy. Fossil fuels can only be used once before their energy is gone. The wind will never run out because the Sun makes the wind, and the Sun will not run out of its energy for billions of years. Wind energy does not damage the environment in the way that burning fossil fuels does. Harnessing the wind is not dangerous, dirty, hard work like mining coal or drilling for oil and gas. Wind technology is becoming cheaper. Already, electricity from the latest wind turbines is as cheap as electricity from a power station that burns fossil fuels.

Power from sea and mountain winds

In the future, some wind farms may be built at sea. This would solve the problems of noise and ugliness. However, the sea wind carries salt spray, which could damage the moving parts of a wind turbine by eating away the metal. This would mean that sea wind turbines would need to be replaced more often than land turbines. In Denmark, an experimental wind farm was opened in 1991 near a small island off the coast. Eleven wind turbines were built in the sea. If more wind farms like this are built over the next 30 years, nearly half of Denmark's electricity will be made using wind power. Other wind farms are being planned for the British coast. As wind technology becomes more reliable, it might be possible to build some wind farms in very remote areas such as high mountains. If the wind turbines could be operated by remote control, nobody would need to look after them. The turbines would be out of sight and there would be no noise problems.

A wind farm in Le Nordais, Canada.

Gigantic fans are used in a wind tunnel.

Wind power in science

To learn how to use wind power for the future, scientists need to understand how different winds behave and the effects they have on the equipment that uses wind energy. To do this, they have built **laboratories** where they can create winds of any strength they need for their experiments. These places are called wind tunnels. A wind tunnel is usually a long room with a machine at one end which is used for blowing air. Wind tunnels are used to test the effects of wind on windmill blades, sails, and other machines that run on wind power.

Scientists use wind tunnels to learn which shapes move fastest and most easily through the air. They have discovered that rounded shapes move through the air more easily than square ones. Experiments like this have helped scientists to design cars, trucks, and trains that have curved shapes with smoother edges so that the air flows easily past them. Vehicles like this use less fuel. This helps lower costs and cut pollution until better sources of clean, renewable energy are more readily available.

Fact file

Not all experiments using wind power were to test wind-powered equipment. Scientists and inventors such as Benjamin Franklin and Alexander Graham Bell flew kites in simple tests to understand what electricity was and how it worked.

Advantages and disadvantages of wind power

Fossil fuels are non-renewable sources of energy. If we keep using them at the current rate,
- coal will run out in 250 years
- oil will run out in 90 years
- gas will run out in 60 years.

There are other sources of energy that are cleaner, safer, and will not run out. Wind energy is a safe, clean, and renewable source of energy suitable for the world's future power needs.

Wind turbines make electricity without polluting the environment.

ADVANTAGES OF WIND ENERGY	DISADVANTAGES OF WIND ENERGY
• Wind energy does not cause pollution.	• On some days the wind does not blow, so a backup source of electricity is always needed.
• Wind energy is useful for remote places. It can be tailored for use in individual households (using, for example, individual wind turbines, wind-turbine kits, and homemade wind turbines) as well as for larger areas.	• A lot of space is needed for large wind farms to generate electricity.
• Wind-turbine technology is becoming cheaper. This makes wind energy available as a choice for more and more people.	• Some people think that wind farms are ugly and that wind turbines are too noisy and a danger to birds.
• The wind is free and will not run out.	

Glossary

acid a type of chemical that can be harmful to people and the environment

airborne in the air

axis an imaginary line through the center of a thing that rotates

batteries containers filled with chemicals that can store or produce electricity

crops plants grown for food

developing countries countries that are beginning to use modern technology

efficient without waste

environmentalists people who care for the environment

fiberglass a material made from thin threads of glass

flexible able to bend

fuel cells devices that produce electricity from a continuous chemical reaction inside them

gasoline a liquid fuel made from oil that is burned inside an engine

generator a machine that turns energy into electricity

global warming a rise in the temperature of the Earth's surface caused by burning fossil fuels

harnessed to be controlled and put to work

horizontal level with the horizon

laboratories places where scientists carry out tests

magnet a piece of iron which makes steel or other metal move towards it

nacelle the control room inside the top part of a wind turbine, behind the blades, which contains the computer

remote very far away from other people

rudder a wide, flat board attached to the end of a boat which is used for steering

synthetic made by humans

trade buying, selling, or swapping something

turbines motors with wheels that spin when they are pushed by a stream of air, water, or steam

vertical upright

wells holes drilled into the Earth

Index

A
air-cushion vehicles 24
airfoils 20–21, 22, 26

D
Darrieus rotor 14

E
electricity 7, 9, 11, 12, 15, 28, 29

F
fossil fuels 4, 7, 9, 11, 15, 18, 19, 24, 28, 30

G
gliders 22
global warming 17

H
hang-gliders 22
horizontal windmills 10
hot-air balloons 22
hovercraft 24

I
ice-yachts 27

K
kites 8, 23, 25, 29

L
land-yachts 27

N
non-renewable energy 4, 11, 15, 30

P
pollution 7, 17, 30
post windmills 10

Q
quadrifoils 23

R
renewable energy 4, 5, 11, 17, 28, 30

S
sailing ships 7, 8, 9, 19, 20, 21
steam power 9

T
tower windmills 11

W
water pumps 7, 16, 18
wind, how it is made 6
wind farms 15, 17, 28, 30
wind-skating 26
wind tunnels 29
wind turbines 13–14, 15, 16, 17, 18, 28, 30
windmills 7, 9, 10–12
windsurfing 26
wing sails 21

Y
yachts 20, 27